Informing the legislative debate since 1914

The National Science Foundation: Background and Selected Policy Issues

Heather B. Gonzalez
Specialist in Science and Technology Policy

June 5, 2014

Congressional Research Service

7-5700

www.crs.gov

R43585

CRS REPORT
Prepared for Members and
Committees of Congress

Summary

The National Science Foundation (NSF) supports both basic research and education in the non-medical sciences and engineering. Congress established the foundation as an independent federal agency in 1950 and directed it to "promote the progress of science; to advance the national health, prosperity, and welfare; to secure the national defense; and for other purposes." The NSF is a primary source of federal support for U.S. university research, especially in certain fields such as mathematics and computer science. It is also responsible for significant shares of the federal science, technology, engineering, and mathematics (STEM) education program portfolio and federal STEM student aid and support.

The NSF is an independent federal agency. Although governed by the budget and oversight processes, NSF's independent status has provided it with greater institutional autonomy than some other federal agencies. Some analysts assert that autonomy protects NSF's scientific mission. However, it may also be perceived as existing in tension with other public values, such as accountability. The tension between independence and accountability is an enduring policy theme for the NSF. It is reflected in debates over the foundation's authorization period and the role (if any) of Congress in grant-making and research prioritization.

NSF is the second-largest source of federal funding for basic research. Since FY2007, increases in the NSF budget have been driven by the doubling policy for physical sciences and engineering (PS&E) research. The PS&E doubling policy sought to double funding for targeted accounts at the National Institute of Standards and Technology, the Department of Energy's Office of Science, and NSF (total). Although this policy was authorized and reauthorized in the America COMPETES Act (P.L. 110-69) and America COMPETES Reauthorization Act of 2010 (P.L. 111-358), and was pursued by both the George W. Bush and Obama Administrations, actual appropriations have not met authorized levels in most fiscal years. (Funding increased, just not to authorized levels.) PS&E doubling provisions expired in FY2013. President Obama did not seek funding for a PS&E doubling in FY2014 or FY2015. Bills to reauthorize NSF and other targeted accounts in the 113th Congress (specifically, H.R. 4186 and H.R. 4159) would authorize funding for some or all of the targeted accounts at growth rates that are below the paces set by the COMPETES acts.

In addition to its research responsibilities, NSF is the only federal agency whose primary mission includes education across all fields of science and engineering. (In this respect, NSF has a dual mission: research and education.) Funding for STEM education activities at NSF typically constitutes about a third of the total federal STEM education effort. Key questions for the 113th Congress focus on the Obama Administration's overall effort to reorganize the federal STEM education effort and the consequences of those changes for STEM education programs at NSF; the direction of the overarching federal STEM education strategy and NSF's role therein; as well as funding for STEM education at the foundation, as a percentage of total NSF appropriations.

NSF received $7.172 billion in funding (estimated) in FY2014. Typically, about 80% of the NSF budget supports the main research account, 12% supports the main education account, 3% to 5% supports facilities and construction, and the remainder supports administrative and related activities. Since FY2006, NSF appropriations have usually been included in annual Commerce, Justice, Science and Related Agencies appropriations acts. Major NSF authorizations expired in FY2013. At least two bills to reauthorize the foundation have been introduced in the 113th Congress (H.R. 4186 and H.R. 4159).

Contents

Introduction 1
Structural Characteristics 1
Leadership and Staff 2
Mission Activities 4
Grant-Making 9
Scientific Facilities, Instruments, and Equipment 11
Major Constituencies 12
Selected Authorizations 13
    Legislative Origin 13
    America COMPETES 17
    Reauthorization Activity in the 113th Congress 22
Budget and Appropriations 22
    FY2015 22
    FY2014 23
    FY2013 24
Concluding Observations 24

Figures

Figure 1. Distribution of Funding for NSF Mission Activities 5

Tables

Table 1. NSF Appropriations by Decade: FY1951 to FY2010 19
Table A-1. Selected NSF Authorizations Acts 26
Table B-1. NSF Authorizations, Budget Requests, and Appropriations 27
Table B-2. NSF Obligations by Major Account: FY2003-FY2013 29

Appendixes

Appendix A. NSF Authorizations Acts 26
Appendix B. NSF Funding History 27

Contacts

Author Contact Information 30

Introduction

The National Science Foundation (NSF) supports both basic research and education in the non-medical sciences and engineering. Congress established the foundation as an independent federal agency in 1950 and directed it to "promote the progress of science; to advance the national health, prosperity, and welfare; to secure the national defense; and for other purposes."[1] The NSF is a primary source of federal support for U.S. university research, especially in certain fields such as mathematics and computer science. It is also responsible for significant shares of the federal science, technology, engineering, and mathematics (STEM) education program portfolio and federal STEM student aid and support.

This report includes information about the NSF for readers seeking an introduction to the foundation and its work. It is intended to provide background and institutional context for ongoing congressional consideration of NSF policy and fiscal issues.

Structural Characteristics

Certain NSF structural characteristics set the foundation apart from other federal agencies and strongly influence its relationship with Congress. In particular, inventories of various federal agencies classify the NSF as an "independent agency." Two of the characteristics that contribute to this classification include NSF's position within the executive branch—it is freestanding, not within an executive department—and its leadership arrangement.[2] The NSF (unlike many other federal agencies) is governed by a 24-member board and a director, each of whom are appointed by the President to fixed, six-year terms.[3] Further, the foundation's organic act specifically establishes it as an "independent agency."[4] This independence, however, is not absolute. For example, the NSF's authorizing statute expressly references the President's authority to remove the director. Further, both Congress and the President retain the power to govern the NSF

> **A Central Tension**
>
> In varying ways and to varying degrees, Congress has grappled with the tension between scientific independence and public accountability at the NSF since the foundation was established in 1950. (See section on "Legislative Origin.") This tension has remained a central policy theme for the NSF throughout its history. It is embedded in the very nature of the NSF as a federal entity and it underpins a wide variety of NSF policy debates—such as the debate about the length of foundation authorization periods (three years? five years? one?) and the debate about the NSF's grant-making process and merit-review criteria, which Congress has sought to influence on a number of occasions. Some policy makers assert that the foundation can best accomplish its scientific purposes if free from undue political influence, while others seek to ensure accountability in the expenditure of public funds. Each Congress has the opportunity to revisit this tension and to redefine the relationship between the NSF and Congress.

---

[1] P.L. 81-507.

[2] See David E. Lewis and Jennifer L. Selin, *Sourcebook of United States Executive Agencies*, 1st ed. (Washington, DC: Administrative Conference of the United States, March 2013), p. 54; J. Merton England, "National Science Foundation," in *Government Agencies*, ed. Donald R. Whitnah (Westport, CT: Greenwood Press, 1983), pp. 367-372; and Harold Seidman, "A Typology of Government," in *Federal Reorganization: What Have We Learned*, ed. Peter Szanton (Chatham, NJ: Chatham House Publishers, Inc. 1981), pp. 43-44.

[3] The NSF director must also be confirmed by the Senate.

[4] 42 U.S.C. 1861, §2.

through the budget, appropriations, and oversight processes.

Policy makers express a variety of rationales for establishing independent agencies, including the belief that independence will facilitate better decision-making (particularly with respect to complex, ostensibly apolitical technical issues) or the desire to free agencies from the control and direction of the executive.[5] In the NSF's case, one NSF historian has observed, "Although the director was subject to removal by the President, his six-year statutory term, like that of the board members, *showed a desire to insulate the agency from politics* [emphasis added]."[6] Some analysts find trade-offs to agency independence though, noting that (in general), "autonomy can be a means of helping [agencies] accomplish democratic purposes ... however, [it] also shields them from direct accountability."[7] As a practical matter, legislators seeking to apply various federal assets toward specific national goals may find both benefits and barriers in the foundation's status as an independent agency.

Leadership and Staff

Consistent with the foundation's purposes, NSF leadership and staff include highly trained scientists and engineers from across the United States. More than half of NSF employees have earned at least a master's degree, over a quarter have a doctorate, and about 15% have completed post-doctorate education. In FY2012, about three-quarters of NSF staff held permanent appointments and about a quarter held non-permanent positions.

Leadership. The National Science Foundation is governed jointly by the NSF director and the 24-person National Science Board (NSB).[8] The director oversees the day-to-day activities of the foundation, including staff and management, program creation and administration, grant-making and merit review, planning, budget, and operations.[9] The board identifies issues critical to NSF's future, approves the foundation's strategic budget direction, approves annual budget submissions to the Office of Management and Budget, ensures balance between initiatives and core programs, and approves new major programs and awards.[10] The board also serves as an independent body of advisors to Congress and the President. NSF's director is an *ex officio* member of the board. NSB members typically come from industry and academia and represent a variety of STEM

---

[5] For more information about independent agencies—including rationales for, historical origin of, and accountability in—see the section titled, "Background and Context" in CRS Report R43391, *Independence of Federal Financial Regulators*, by Henry B. Hogue, Marc Labonte, and Baird Webel.

[6] England, p. 367.

[7] Lewis and Selin, p. 59.

[8] More information about NSF leadership and staff may be found in Stephen Horn (panel chair), et al., *National Science Foundation: Governance and Management for the Future*, National Academy of Public Administration, April 2004, p. xv, at http://www.napawash.org/2004/1539-national-science-foundation-governance-and-management-for-the-future.html.

[9] National Science Foundation, "About the National Science Foundation: Who We Are," *National Science Foundation Website*, accessed February 7, 2014, at http://www.nsf.gov/about/who.jsp.

[10] National Science Board, "About the NSB," *National Science Board Website*, accessed February 7, 2014, http://www.nsf.gov/nsb/about/; and National Science Foundation, "Introduction," *Proposal and Award Policies and Procedures Guide* (NSF 13-1), January 13, 2013, at http://www.nsf.gov/pubs/policydocs/pappguide/nsf13001/index.jsp. See also, the section titled "Grant-Making" in this report.

disciplines.[11] Historically, most NSF directors have come from the fields of physics or engineering.[12]

Appointment and Terms of Office. The President appoints the NSF director with the advice and consent of the Senate. The President also appoints the members of the National Science Board. (In 2012 Congress enacted legislation removing Senate confirmation requirements for the members of the NSB.)[13] Both the NSF director and members of the National Science Board serve six-year terms. NSB terms are staggered such that one-third of the board is appointed every two years.

Deputy Director. 42 U.S.C. 1864a provides statutory authority for the NSF-wide deputy director and provides the deputy director with the power to act as NSF director in the event of a vacancy, disability, or absence. The deputy director also performs other duties as determined by the director.[14] Since the mid-1990s the deputy director has served as NSF's Chief Operating Officer. The President appoints the NSF deputy director with the advice and consent of the Senate. The position includes no statutorily prescribed term of office.

Assistant Directors. The leaders of NSF's directorates carry the title "assistant director." The assistant director position is not currently statutorily authorized, but it has been in the past. In FY2014, there were eight assistant directors. Assistant director duties vary by directorate and in some cases have changed over the years. In general, assistant directors lead directorate programs and initiatives and are responsible for planning and implementing programs, priorities, and policies. Assistant directors are often non-permanent staff. In previous years, this position required presidential appointment and Senate confirmation.

Division Directors. Division directors are responsible for long-range planning and budgetary stewardship within their research areas. They also oversee the grant-making process and, in many cases, make the final programmatic decision to approve (or decline) awards to NSF grant-seekers.

Program Directors. Program directors are subject matter experts. They conduct the scientific, technical, and programmatic review and evaluation of proposals, including peer reviewer recruitment and management of the proposal review process. They manage program budgets and provide award oversight. Program directors make funding recommendations to division directors.

Rotators. The NSF workforce is made up of permanent, temporary, and "rotating" staff. Unlike permanent and temporary staff, most rotating staff are hired under the authority of the Intergovernmental Personnel Act of 1970 (IPA, P.L. 91-648) and are not considered federal employees. IPA rotators typically come from institutions of higher education, but they may come from other organizations as well (e.g., state and local government, Indian tribal government, non-profit entities). IPA rotators retain ties to their home institutions—including pay and benefits—and may serve the NSF for no more than four years. Unlike its practices for permanent and temporary staff, NSF uses program funds to provide salary reimbursement, living expenses, and

---

[11] Board members must be "eminent in the fields of the basic, medical, or social sciences, engineering, agriculture, education, research management or public affairs...." (42 U.S.C. 1863(c)(1)).

[12] National Science Foundation, "List of NSF Directors, 1950-Present," *National Science Foundation Website*, accessed February 7, 2013, at http://www.nsf.gov/od/nsf-director-list/nsf-directors.jsp.

[13] Presidential Appointment Efficiency and Streamlining Act of 2011 (P.L. 112-166, §2 (s)).

[14] A list of NSF deputy directors is available at http://www.nsf.gov/od/nsf-director-list/nsf-deputys.jsp.

travel expenses for IPA rotators. Overall, IPA rotators comprised 12% of NSF's total workforce in August 2012. Further, of the 512 program directors at NSF in FY2012, 262 (51%) were permanent employees, 172 (34%) were IPA rotators, 39 (8%) were temporary employees, and 39 (8%) were visiting scholars.

> **NSF Rotators: Pros and Cons**
>
> Policy analysts debate NSF's use of rotators.[15] NSF asserts that rotators bring fresh, cutting-edge insight to foundation programs and that rotators increase knowledge transfer between the research community and the foundation. A 2004 National Academy of Public Administration panel largely substantiated these claims.[16] But NSF's Inspector General notes that IPA rotators are more expensive than federal employees and contends that NSF could enhance oversight and management of its rotator programs.[17]

Mission Activities

NSF's dual mission is to support basic research[18] and education in the non-medical sciences and engineering. NSF is the second-largest source of federal funding for basic research, and a top three source of federal funding for research in the fields of environmental sciences, life sciences, mathematics and computer sciences, physical sciences, social sciences, and other sciences.[19] Funding for STEM education activities at NSF typically constitutes about a third of the total federal STEM education effort.

The foundation currently divides its mission activities among seven directorates, which are mainly organized by academic discipline.[20] Until FY2013, the largest directorate (measured by budget authority) was Mathematical and Physical Sciences (MPS, $1.312 billion in FY2011 actual) and the smallest was Social, Behavioral, and Economic Sciences (SBE, $247.3 billion in

---

[15] For example, see Jeffrey Mervis, "Special Report: Can NSF Put the Right Spin on Rotators? Part 1," *Science Insider*, October 10, 2013, http://news.sciencemag.org/policy/2013/10/special-report-can-nsf-put-right-spin-rotators-part-1; and Jeffrey Mervis, "Special Report: Can NSF Put the Right Spin on Rotators? Part 2," *Science Insider*, October 24, 2013, at http://news.sciencemag.org/people-events/2013/10/special-report-can-nsf-put-right-spin-rotators-part-2.

[16] Stephen Horn (panel chair), et al., *National Science Foundation: Governance and Management for the Future*, National Academy of Public Administration, April 2004, p. xv, at http://www.napawash.org/2004/1539-national-science-foundation-governance-and-management-for-the-future.html.

[17] Memorandum from Assistant Inspector General for Audit Dr. Brett M. Baker, National Science Foundation, Office of Inspector General to Deputy Director Dr. Cora B. Marrett, National Science Foundation, dated March 20, 2013, at http://www.nsf.gov/oig/13-2-006.pdf.

[18] OMB Circular A-11, Schedule C, defines basic research as "systematic study directed toward fuller knowledge or understanding of the fundamental aspects of phenomena and of observable facts without specific applications towards processes or products in mind. Basic research, however, may include activities with broad applications in mind." Basic research differs from applied research, which is "systematic study to gain knowledge or understanding necessary to determine the means by which a recognized and specific need may be met;" and from development, which is the "systematic application of knowledge or understanding, directed toward the production of useful materials, devices, and systems or methods, including design, development, and improvement of prototypes and new processes to meet specific requirements." See, Office of Management and Budget, "Character Classification (Schedule C)," *OMB Circular A-11* (2013), at http://www.whitehouse.gov/sites/default/files/omb/assets/a11_current_year/s84.pdf.

[19] Based on preliminary FY2012 data from Tables 29 and 22 of the National Science Foundation, National Center for Science and Engineering Statistics, Federal Funds for Research and Development: Fiscal Years 2010-12, NSF 13-326 (July 2013).

[20] NSF's organizational chart is available at http://www.nsf.gov/staff/orglist.jsp. In addition to the research and education directorates, NSF also has two administrative offices: the Office of Budget, Finance, & Award Management and the Office of Information & Resource Management.

FY2011 actual). NSF adjusted its organization chart in FY2013, placing some previously free-standing offices into existing directorates. As a result of these changes, the Directorate for Geosciences became the largest directorate at NSF. SBE remained the smallest. (See **Figure 1**.)

Figure 1. Distribution of Funding for NSF Mission Activities
FY2013 Actual, by Directorate

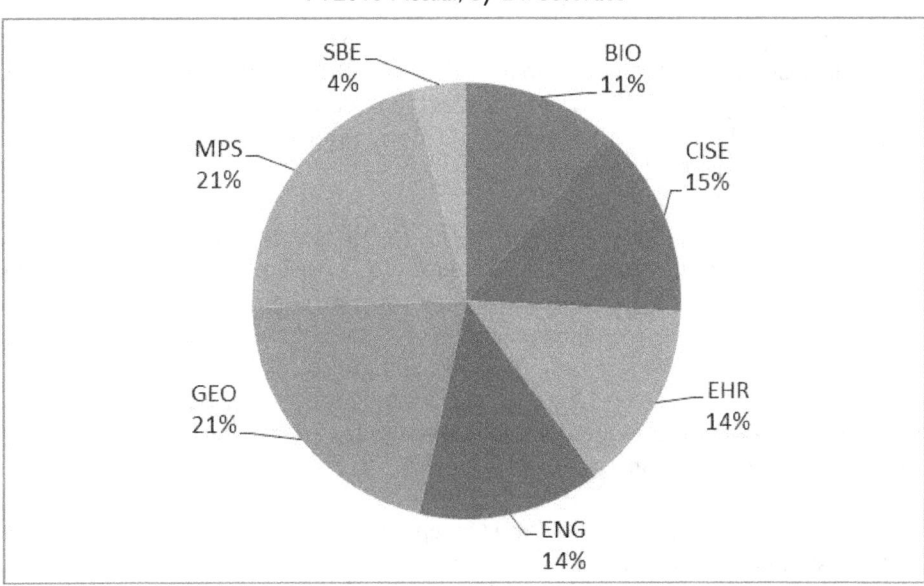

Source: Congressional Research Service, based on FY2013 actual funding levels reported in NSF's *FY2015 Budget Request to Congress*, available at http://www.nsf.gov/about/budget/fy2015/toc.jsp.

Notes: The term "SBE" refers to the Directorate for Social, Behavioral, and Economic Sciences; "BIO" refers to the Directorate for Biological Sciences; "CISE" refers to the Directorate for Computer and Information Science and Engineering; "EHR" refers to the Directorate for Education and Human Resources; "ENG" refers to the Directorate for Engineering; "GEO" refers to the Directorate for Geosciences; and "MPS" refers to the Directorate for Mathematical and Physical Sciences.

NSF directorates are divided into divisions—with between four and six divisions per directorate, typically—that manage programs. The main appropriations account for all but one directorate is NSF's Research and Related Activities (RRA) account. The Directorate for Education and Human Resources is the exception to this rule; its main source of appropriations is the Education and Human Resources (EHR) account. Many NSF programs or projects are co-funded (i.e., they receive funding from two or more foundation accounts) or involve coordination and cooperation between programs and disciplines.

Directorate for Biological Sciences (BIO). BIO's mission "is to enable discoveries for understanding life. BIO-supported research advances the frontiers of biological knowledge, increases our understanding of complex systems, and provides a theoretical basis for original research in many other scientific disciplines."[21] BIO divisions include Molecular and Cellular

---

[21] National Science Foundation, Directorate for Biological Sciences, "About Biological Sciences," *National Science Foundation Website*, accessed February 11, 2014, at http://www.nsf.gov/bio/about.jsp. See also, National Science Foundation, *Understanding Life: Biological Sciences*, NSF 14-804 (March 2014), at https://www.nsf.gov/about/congress/reports/bio_research.pdf.

Biosciences; Integrative Organismal Systems; Environmental Biology, Biological Infrastructure; and Emerging Frontiers. In FY2013, BIO received $679 million in funding (actual).

Directorate for Computer and Information Science and Engineering (CISE). CISE "supports investigator-initiated research in all areas of computer and information science and engineering, fosters broad interdisciplinary collaboration, helps develop and maintain cutting-edge national computing and information infrastructure for research and education, and contributes to the development of a computer and information technology workforce with skills essential for success in the increasingly competitive global market."[22] CISE divisions include Advanced Cyberinfrastructure, Computing and Communication Foundations, Computer and Network Systems, Information and Intelligent Systems, and Information Technology Research. FY2013 actual funding for CISE was $858 million.

Directorate for Education and Human Resources (EHR). EHR seeks to "achieve excellence in U.S. science, technology, engineering and mathematics (STEM) education at all levels and in all settings (both formal and informal) in order to support the development of a diverse and well-prepared workforce of scientists, technicians, engineers, mathematicians and educators and a well-informed citizenry that have access to the ideas and tools of science and engineering."[23] EHR has four divisions: Research on Learning in Formal and Informal Settings, Graduate Education, Human Resource Development, and Undergraduate Education. FY2013 actual funding for EHR was $835 million.

Directorate for Engineering (ENG). ENG "investments in engineering research and education aim to build and strengthen a national capacity for innovation that can lead over time to the creation of new shared wealth and a better quality of life."[24] ENG divisions include Chemical, Bioengineering, Environmental, and Transport Systems; Civil, Mechanical, and Manufacturing Innovation; Electrical, Communications, and Cyber Systems; Engineering Education and Centers; Industrial Innovation and Partnerships; and Emerging Frontiers in Research and Innovation. FY2013 actual funding for ENG was $820 million.

Directorate for Geosciences (GEO). GEO's mission is "to support research in the atmospheric, earth, and ocean sciences."[25] GEO divisions include Atmospheric and Geospace Sciences, Earth Sciences, Integrative and Collaborative Education and Research, Ocean Sciences, and Polar Programs. FY2013 actual funding for GEO was $1.250 billion.

---

[22] National Science Foundation, Directorate for Computer and Information Science and Engineering, "CISE—About," *National Science Foundation Website*, accessed February 11, 2014, at http://www nsf.gov/cise/about.jsp. See also, National Science Foundation, *Enhancing Our Lives through Computing: Computer & Information Science & Engineering*, NSF 14-807 (February 2014), at https://www nsf.gov/about/congress/reports/cise_research.pdf.

[23] National Science Foundation, Directorate for Education and Human Resources, "About Education and Human Resources," *National Science Foundation Website*, accessed February 11, 2014, at http://www nsf.gov/ehr/about.jsp. See also, National Science Foundation, *Inspiring STEM Learning: Education & Human Resources*, NSF 12-800 (September 2013), at https://www.nsf.gov/about/congress/reports/ehr_research.pdf.

[24] National Science Foundation, Directorate for Engineering, "General Information about ENG," *National Science Foundation Website*, accessed March 25, 2014, at http://www.nsf.gov/eng/about.jsp. See also, National Science Foundation, *Making Future Technologies Possible: Engineering*, NSF14-808 (April 2014), at https://www.nsf.gov/about/congress/reports/eng_research.pdf.

[25] National Science Foundation, Directorate for Geosciences, "About GEO," *National Science Foundation Website*, accessed February 11, 2014, at http://www nsf.gov/geo/about.jsp. See also, National Science Foundation, *Unraveling Earth's Complexity: Geoscience*s, NSF 13-801 (September 2013), at https://www nsf.gov/about/congress/reports/geo_research.pdf.

Directorate for Mathematical and Physical Sciences (MPS). MPS's mission is to "harness the collective efforts of the mathematical and physical sciences communities to address the most compelling scientific questions, educate the future advanced high-tech workforce, and promote discoveries to meet the needs of the Nation."[26] MPS divisions include Astronomical Sciences, Chemistry, Materials Research, Mathematical Sciences, and Physics. FY2013 actual funding for MPS was $1.274 billion.

Directorate for Social, Behavioral, and Economic Sciences (SBE). SBE's mission is "to promote the understanding of people and their lives by supporting research that reveals basic facets of human behavior; to encourage research that addresses important societal questions and problems; to work with other scientific disciplines to ensure that basic research and solutions to problems build upon the best multidisciplinary science; and to provide mission-critical statistical information about science and engineering (S&E) in the U.S. and the world through the National Center for Science and Engineering Statistics (NCSES)."[27] In addition to the NCSES, SBE divisions include Social and Economic Sciences, as well as Behavioral and Cognitive Sciences. FY2013 actual funding for SBE was $243 million.

---

[26] National Science Foundation, Directorate for Mathematical and Physical Sciences, "About Directorate for Mathematical and Physical Sciences," *National Science Foundation Website*, accessed February 11, 2014, at http://www.nsf.gov/mps/about.jsp. See also, National Science Foundation, *Enriching the Language of Discovery: Mathematical & Physical Sciences*, NSF 14-805 (March 2014), https://www.nsf.gov/about/congress/reports/mps_research.pdf.

[27] National Science Foundation, *FY2015 Budget Request to Congress*, March 10, 2014, p. SBE-1. See also, National Science Foundation, Exploring What Makes Us Human: Social, Behavioral & Economic Sciences, NSF 14-803 (March 2014), at https://www.nsf.gov/about/congress/reports/sbe_research_2.pdf.

> ### Funding for Social Science?
>
> Congress has grappled with the question of funding for the social sciences since NSF was established. The debate generally centers on the question of public benefit. Those who seek to reduce funding for research in the social, behavioral, and economic (SBE) sciences typically assert that federal dollars should focus on fields perceived to be more closely associated with national security, health, or economic interests, such as the physical and life sciences. Some opponents also question certain NSF SBE grants—such as the "study of human-set forest fires 2,000 years ago in New Zealand"—which may be perceived as frivolous or wasteful.[28] Supporters, on the other hand, typically dispute the notion that SBE research does not serve the national interest, citing research that is perceived to provide broad public benefit, such as research to improve disaster response, facilitate kidney matching between donors and patients, and research that informs innovation.[29] Supporters maintain that SBE grants can yield critical innovations (though they may appear frivolous) or assert that NSF's peer-review process assures that only meritorious proposals are funded.
>
> The floor debate over H.Amdt. 734—which amended H.R. 4660 (Commerce, Justice, Science and Related Agencies Appropriations Act, FY2015) to reduce funding for NSF's SBE directorate by $15.4 million and to redirect the funds to physical sciences and engineering—offers an example of this policy debate in the 113th Congress.[30] Similar debates have occurred in the context of deliberations over H.R. 4186 (Frontiers in Innovation, Research, Science, and Technology Act of 2014), which (among other things) seeks to prioritize funding for physical sciences and engineering research at the NSF. H.R. 4186 was reported favorably from the House Committee on Science, Space, and Technology on May 28, 2014.
>
> Legislators have also focused more narrowly on the specific question of funding for NSF's Political Science program, which is part of SBE. Some legislators assert that political science research is extraneous to NSF's central mission and a waste of federal dollars.[31] Legislators who hold this view introduced several provisions limiting funding for political science at NSF in the 112th and 113th Congresses. Most of these provisions have not been enacted. However, one provision limiting funding for NSF's Political Science program in FY2013 (Section 543 of P.L. 113-6, Consolidated and Further Continuing Appropriations Act, 2013) became law.[32]
>
> Some of those who objected to Section 543 asserted that it threatened NSF's independence and therefore the integrity of the research the foundation supports.[33] They also contended that Section 543 put the decision about what research to fund in the hands of Congress rather than in the hands of the scientists who lead and manage NSF (and who are, they argue, better able to judge the intellectual merit of various research proposals). Similar provisions limiting funding for political science were not enacted in FY2014. Average annual funding for NSF's Political Science program is typically in the $9 million to $10 million range.

---

[28] Rep. Lamar Smith, "Commerce, Justice, Science, and Related Agencies Appropriations Act, 2015," remarks in the House, *Congressional Record*, daily edition, vol. 160 (May 29, 2014), pp. H4958.

[29] National Science Foundation, *Bringing People into Focus: How Social, Behavioral and Economic Research Addresses National Challenges*, at https://www.nsf.gov/about/congress/reports/sbe_research.pdf. See also, Jameson M. Wetmore, "The Value of the Social Sciences for Maximizing Public Benefits of Engineering," *The Bridge: Linking Engineering and Society*, vol. 42, no. 3 (Fall 2012), p. 41.

[30] Smith, ibid., pp. H4957-H4960.

[31] For example, see Sen. Tom Coburn, M.D., "Dr. Coburn Offers Amendments to 587-Page, $1 Trillion Continuing Resolution to Fund the Government," press release, March 13, 2103, at http://www.coburn.senate.gov/public/index.cfm/rightnow?ContentRecord_id=1167a7ba-31a0-4873-b5c9-a2acd77eb72b.

[32] As adopted, Section 543 prohibited NSF from using FY2013 appropriations to carry out the functions of the Political Science Program, except for projects that promote U.S. national security and economic interests. Section 543 further required the NSF to certify that political science projects met this requirement, to publish the reason for each certification on its website, and allowed NSF to use any unobligated balances from the Political Science Program for other research.

[33] For example, see American Political Science Association, "Senate Delivers a Devastating Blow to the Integrity of the Scientific Process at the National Science Foundation," press release, March 20, 2013, at http://www.prnewswire.com/news-releases/senate-delivers-a-devastating-blow-to-the-integrity-of-the-scientific-process-at-the-national-science-foundation-199221111.html.

Grant-Making

Because most NSF funding is distributed to researchers and institutions outside of the foundation, grant-making is (arguably) the heart of what NSF does.[34] Grants can be either standard (i.e., full funding up-front) or continuing (i.e., incremental funding on a multi-year basis). NSF receives approximately 50,000 grant proposals annually.[35] Of these, between 20% and 25% typically receive funding. About 36,500 scientists and engineers participated in the merit review process as panelists and proposal reviewers in FY2013.

The vast majority of NSF grants are awarded through a competitive, merit-based assessment process.[36] The peer review stage of this process—in which external "peer" reviewers with subject matter expertise assess the merits of each grant proposal—is both widely lauded and closely watched by policy analysts. Although peer review is perhaps the most well-known stage of NSF's grant-making process, peer review does not encompass the whole of the assessment process. Rather, the typical grant-making process for most NSF awards follows three phases.

- Phase 1: opportunity announced, proposals submitted, proposals received.[37]
- Phase 2: reviewers selected, peer review, program officer recommendation, division director review.[38]
- Phase 3: business review, award finalized.[39]

Put differently, most NSF proposals must survive at least five kinds of scrutiny. First, the initial assessment is for application completeness and conformance with NSF requirements. Second, if a proposal survives the initial assessment, then it is sent to three or more external subject matter experts for peer review.[40] Peer reviewers evaluate the proposal according to two broad criteria: intellectual merit and broader impacts.[41] According to the NSF,

> "*Intellectual Merit:* The Intellectual Merit criterion encompasses the potential to advance knowledge; and

---

[34] NSF's *Proposal and Award Policies and Procedures Guide* (PAPPG) describes the foundation's grant-making process and provides guidance to potential applicants.

[35] PAPPG, "Introduction."

[36] One exception to this rule is the RAPID (Grants for Rapid Response Research) funding mechanism. RAPID grants may be used for "proposals having a severe urgency with regard to availability of, or access to data, facilities or specialized equipment, including quick-response research on natural or anthropogenic disasters and similar unanticipated events." Only internal merit review is required for RAPID grants. PAPPG, p. II-22.

[37] More information on the proposal preparation and submission phase is available at http://www.nsf.gov/bfa/dias/policy/merit_review/phase1.jsp.

[38] More information on the proposal review and processing phase is available at http://www.nsf.gov/bfa/dias/policy/merit_review/phase2.jsp.

[39] More information on the award processing phase is available at http://www.nsf.gov/bfa/dias/policy/merit_review/phase3.jsp.

[40] Peer review can happen in a number of ways. Reviewers may be sought out on an individual basis (also known as ad hoc review) or may participate in in-person or virtual panels. While a minimum of three reviewers is required, more may participate.

[41] In addition to these criteria, NSF solicitations may include additional criteria that meet the specific objectives of programs or activities.

Broader Impacts: The Broader Impacts criterion encompasses the potential to benefit society and contribute to the achievement of specific, desired societal outcomes."[42]

Peer reviewers provide information about the merit of the proposal to the program director, who (third) considers the proposal in the context of the broader program portfolio and direction. Program directors are not bound by the recommendations of peer reviewers. Rather, the program director

> reviews the proposal and analyzes the input received from the external reviewers. In addition to the external reviews, Program [Directors] consider several factors in developing a portfolio of funded projects. For example, these factors might include different approaches to significant research and education questions; potential (with perhaps high risk) for transformational advances in a field; capacity building in a new and promising research area; or achievement of special program objectives. In addition, decisions on a given proposal are made considering both other current proposals and previously funded projects.[43]

Fourth, after the portfolio assessment, the program director submits his or her award recommendation to the division director, who examines the recommendations and typically makes the final programmatic decision to fund (or not).

Fifth, if the proposal survives programmatic review (including initial, peer, program, and division), it is sent to the Office of Budget, Finance, and Award Management (BFA). Analysts within the BFA conduct an assessment of the business, financial, and policy implications, and, if called for, issue the grant.

In addition, larger or "sensitive" awards may require further layers of review beyond those already described, including review by NSF senior management or the National Science Board.[44] This rule applies to all Major Research Equipment and Facilities Construction (MREFC) projects. NSB also establishes average annual award amounts (for each directorate or office) that trigger NSB review and approval requirements. Trigger amounts differ by directorate and ranged from $6.88 million (BIO) to $12.66 million (GEO) in FY2014.[45]

Besides grants, NSF also awards funding though other mechanisms, such as cooperative agreements, contracts, and competitions. In FY2013, NSF distributed 73% of its funding via grants, 22% by cooperative agreement, and 5% by contract.[46]

---

[42] More information about NSF's merit review process and criteria is available at the NSF Merit Review FAQ webpage, available at http://www.nsf.gov/bfa/dias/policy/merit_review/facts.jsp; and in "Chapter III" of the PAPPG, available at http://www.nsf.gov/pubs/policydocs/pappguide/nsf13001/gpg_3.jsp#IIIA2a.

[43] National Science Foundation, "Phase II: Proposal Review and Processing," *National Science Foundation Website*, accessed February 11, 2014, http://www.nsf.gov/bfa/dias/policy/merit_review/phase2.jsp.

[44] The National Science Board resolution, "Delegation of Award-Approval Authority to the Director," (NSB-11-2) establishes the conditions under which the NSB delegates its authority to approve NSF awards to the NSF director. Section (2)(B) states that the director may not make an award that "involves sensitive political or policy issues" without Board approval.

[45] The NSB Delegation of Award-Approval Authority to the Director" (Ibid. Section (1)) states that the NSF director "may make no award involving an anticipated average annual amount of the greater of either 1 percent of more of the awarding Directorate's or Office's prior year current plan *or* [emphasis original] 0.1 percent or more of the prior year total NSF budget without the prior approval of the National Science Board."

[46] Contracts are used to acquire products, services, and studies (e.g., program evaluations).

> **The Geography of Grants**
>
> University research is widely believed to contribute to state and regional economic development. As such, many policy makers seek to increase research funding for their local colleges and universities. However, of the over 900 institutions reporting at least $150,000 in R&D in FY2012, the top 100 accounted for 79% of total academic science and engineering R&D. This trend has held constant for at least two decades. In response, some policy makers have sought mechanisms to reduce concentration in the geographic distribution of federal research grants, which make up a large portion of academic R&D.[47]
>
> To this end, Congress established the Experimental Program to Stimulate Competitive Research (EPSCoR) at NSF in 1978.[48] EPSCoR is NSF's only state-based program. It is designed to strengthen the research capacity of institutions located within "EPSCoR Jurisdictions"—that is, those states that have historically received limited federal R&D funding—so that they are able to compete more successfully for federal R&D funding. Since the NSF EPSCoR program was established at least five other federal agencies have launched EPSCoR programs. The 2014 EPSCoR Jurisdiction map from the EPSCoR Interagency Coordinating Committee, which NSF staffs and chairs, includes between 23 and 31 eligible U.S. states and territories (depending on the agency).[49]
>
> At the beginning of the EPSCoR program, some questioned the length of time required for a state to improve its research infrastructure.[50] A five-year limit was proposed, but that proved to be "... unrealistic, both substantively and politically."[51] Questions remain concerning the length of time states should receive EPSCoR support. Some analysts assert that some states and their institutions should assume more responsibility for building their research infrastructure and become less dependent on EPSCoR funds. They argue that some researchers and states have become comfortable with EPSCoR funding and are not being aggressive in graduating from the program. In three decades no state has graduated from NSF's EPSCoR program. The issue of graduation has generated considerable congressional interest. In 2013, The National Academies published a study of federal EPSCoR programs, including findings and recommendations related to graduation and other program aspects.[52]

Scientific Facilities, Instruments, and Equipment

NSF does not typically directly operate laboratories or scientific facilities. However, the foundation provides operations and maintenance support to a wide array of scientific facilities, instruments, and equipment. NSF funding supports the National Center for Atmospheric Research, polar facilities and logistics, and the academic research fleet. More information about NSF support for scientific facilities, instruments, and equipment is typically found in the "Facilities" sections of NSF's annual budget requests to Congress.

---

[47] "Total academic R&D" includes funding from all sources. However, "the federal government provided the majority of the S&E R&D funds that public and private institutions spent on R&D in FY 2012 (just under 60% and just over 70%, respectively)." National Science Board, *Science and Engineering Indicators 2014*, NSB-14-01 (Arlington, VA: National Science Foundation, 2014), p. 5-16 – 5-17.

[48] Initial funding for EPSCoR was provided in P.L. 95-392 (Department of Housing and Urban Development-Independent Agencies Appropriation Act, 1979). More information about EPSCoR is available at http://www.nsf.gov/div/index.jsp?org=EPSC.

[49] National Science Foundation, EPSCoR Interagency Coordinating Committee, "Eligible EPSCoR Jurisdictions by Agency," fact sheet, last updated May 17, 2012, at http://www.nsf.gov/od/iia/programs/epscor/EICC.pdf.

[50] This paragraph adapted from CRS Report RL30930, *U.S. National Science Foundation: Experimental Program to Stimulate Competitive Research (EPSCoR)*, by Christine M. Matthews (retired).

[51] Lambright, W. Henry, Syracuse University, Paper prepared for the American Association for the Advancement of Science, Workshop on Academic Research Competitiveness, Coeur d'Alene, Idaho, *Building State Science: The EPSCoR Experience*, October 1-3, 1999, at http://www.aaas.org/spp/rcp/epscor/lambright.htm, p.4.

[52] National Academy of Sciences, National Academy of Engineering, and Institute of Medicine, Policy and Global Affairs, Committee on Science, Engineering, and Public Policy, Committee to Evaluate the Experimental Program to Stimulate Competitive Research (EPSCoR) and Similar Federal Agency Programs, *The Experimental Program to Stimulate Competitive Research* (Washington, DC: The National Academies Press, 2013).

NSF's Major Research Equipment and Facilities Construction (MREFC) account also provides funding for (typically) between four and six research facilities and equipment construction projects. These projects include international activities, such as the Atacama Large Millimeter Array (a large radio telescope located in northern Chile that was completed in 2013);[53] as well as U.S. projects, including certain ground-based astronomical telescopes and ecological and ocean observatory networks (which connect geographically distributed scientific facilities and instruments). MREFC funds typically support projects only during the construction phase. Project planning and design, as well as post-construction operations and maintenance, comes from Research and Related Activities (RRA).

Major Constituencies

Approximately three-quarters of NSF funds are awarded annually to colleges, universities, and academic consortia. The remainder of NSF's budget usually goes to private industry (about 13%), Federally Funded Research and Development Centers (around 3%), and other sources (about 3%). NSF typically provides about 50% (or more) of federal funding for academic basic research in computer science, biology, environmental sciences, mathematics, and social sciences. Further, about a third of all known federal funding for STEM education comes from NSF in a typical budget year. The foundation is a primary source of support for graduate student fellowships in the non-biomedical sciences and engineering.

In FY2013, NSF issued 10,844 new awards to 1,922 colleges, universities, and other institutions in 50 states, the District of Columbia, and three U.S. territories. The foundation supported an estimated 299,000 individuals, including researchers, postdoctoral associates, and other professionals; undergraduate and graduate students; and elementary and secondary school teachers and students. At least 212 Nobel laureates have received NSF support at some point in their careers. NSF support for informal science education and scientific literacy reaches many Americans—in museums, libraries, afterschool programs, and through the media—every year.

---

[53] Readers who have seen the 1997 film *Contact* may also be familiar with the NSF-funded Very Large Array in Soccorro, New Mexico. National Science Foundation, "Radio Telescopes in the New Movie 'Contact' Dish Up Real Science," press release (97-049), July 10, 1997, available at http://www.nsf.gov/news/news_summ.jsp?cntn_id=102822.

> **Broadening Participation**
>
> The demographic profile of the U.S. student-age population is changing. The youth population is more racially and ethnically diverse than previous generations of Americans. At the same time, women have attained majority status on many college campuses. Yet, these groups may be underrepresented in certain STEM fields. Some analysts assert that underrepresented groups are an underutilized resource, which could be tapped to help meet perceived demand for STEM-competencies in the U.S. workforce.[54]
>
> General agreement about the problems posed by racial, ethnic, and gender disparities in STEM education and employment has not translated into widespread agreement on either the causes of underrepresentation or policy solutions. Further, causes and solutions may be different for different population subsets. Broadening participation issues include faculty diversity, the potential for bias in grant-making, and "family friendly" work environments for scientists and engineers, as well as teacher quality in schools that serve minority students, parental involvement and support for STEM-interested youth, and access to STEM-related education opportunities and support programs. Broadening participation is not limited to race/ethnicity and gender, either. Studies have shown STEM achievement gaps by income and level of urbanization (e.g., rural, suburban, urban) as well.
>
> NSF operates several dozen programs that seek to broaden participation. The Science and Engineering Equal Opportunities Act (P.L. 96-516), which was incorporated into the 1980 NSF reauthorization, (1) declared that it is U.S. policy to encourage underrepresented populations to participate in STEM, and (2) authorized NSF to establish programs for this purpose. As of FY2013, funding for NSF's broadening participation programs was $755 million (actual). Some of the most widely tracked NSF broadening participation programs provide funding to minority-serving institutions of higher education.

Selected Authorizations

The following sections describe NSF's legislative origins and the foundation's two most recent, enacted reauthorizations as part of the America COMPETES Act (P.L. 110-69) and the America COMPETES Reauthorization Act of 2010 (P.L. 111-358). The 113th Congress has begun the process of reauthorizing certain COMPETES provisions, including NSF provisions. (See section titled, "Reauthorization Activity in the 113th Congress".) **Table A-1** includes a list of selected NSF authorizations dating to the 1950s.

Legislative Origin

Many contemporary policy conversations about the NSF mirror the debate over the foundation's establishment. For example, the 113th Congress has debated the question of funding for social science at the NSF. This issue was also debated during the establishment of the foundation. Retelling the historical conversation, therefore, contextualizes today's deliberations and provides legislators with additional insight into the enduring nature of some of these conflicts. It also provides insight into how previous generations of policy makers resolved similar questions.

Historical accounts of the NSF frequently peg the foundation's genesis to a dialogue between two men: Senator Harley M. Kilgore and Vannevar Bush.[55] Senator Kilgore chaired the Senate

---

[54] National Academy of Sciences, National Academy of Engineering, and Institute of Medicine, Committee on Underrepresented Groups and the Expansion of the Science and Engineering Workforce Pipeline, Committee on Science, Engineering, and Public Policy, Policy and Global Affairs, *Expanding Underrepresented Minority Participation: America's Science and Technology Talent at the Crossroads*, National Academies Press, 2011, at http://www.nap.edu/openbook.php?record_id=12984.

[55] Historical narratives about the founding of the NSF typically focus on Sen. Kilgore and Bush but the 79th Congress considered several bills focused on the question of post-war scientific research. See U.S. Senate, Committee on (continued...)

Subcommittee on War Mobilization during and immediately after World War II. Bush was director of the Office of Scientific Research and Development (ORSD) as well as a science advisor to President Franklin Delano Roosevelt.[56] Between 1942 and 1945, Senator Kilgore's subcommittee held a series of hearings on government support for scientific research. That effort resulted in the July 23, 1945, introduction of S. 1297 (National Science Foundation Act of 1945), which would have established a National Science Foundation. Bush, on the other hand, authored an historic July 1945 report on post-war U.S. scientific research, *Science: An Endless Frontier*, which called for the creation of a National Research Foundation. On July 19, 1945, Senator Warren Magnuson introduced a bill, S. 1285 (National Research Foundation Act of 1945)—which was drafted in consultation with Bush and hewed closely to the proposal outlined in *Science*—to establish a National Research Foundation.[57]

Although Senator Kilgore and Senator Magnuson agreed on the goal of establishing a federal agency for the support of scientific research and their bills shared certain similarities, they promoted different approaches.[58] There was agreement, for example, that the foundation should provide scholarships, that it should support basic research, that it should have both a board and a director, and that it should be independent from other executive branch agencies.[59] Differences focused on five broad themes that would be very familiar to an NSF observer today. These include

- ownership of patents resulting from government research,
- inclusion of the social sciences,
- geographic distribution of funding,
- the extent to which the foundation should support applied research, and
- political and administrative control of the foundation.[60]

(...continued)

Military Affairs, Subcommittee on War Mobilization, *Legislative Proposals for the Promotion of Science: The Texts of Five Bills and Excerpts from Reports*, subcommittee print, 79th Cong., 1st sess., August 1945.

[56] President Franklin Delano Roosevelt established the OSRD as an independent agency within the Office of Emergency Management (Executive Order 8807). More information about OSRD, as well as holdings, is available on the Library of Congress website at http://www.loc.gov/rr/scitech/trs/trsosrd.html.

[57] U.S. Congress, House Committee on Science and Technology, Task Force on Science Policy, "A History of Science Policy in the United States, 1940-1985," *Science Policy Study Background Report No. 1*, 99th Cong., 2nd sess., September 1986 (Washington, DC: GPO, 1986), p. 21-27; and, George T. Mazuzan, *National Science Foundation: A Brief History*, NSF 88-16 (Washington, DC: National Science Foundation, 1988).

[58] At the time, most stakeholders agreed with the general concept of a publically funded scientific research foundation. One exception was Frank B. Jewett, then president of the National Academy of Sciences. Jewett expressed concern about unwanted government control and interference in science and preferred private sources of funding. See J. Merton England, *A Patron for Pure Science: The National Science Foundation's Formative Years, 1945-57* (Washington, DC: NSF, 1982), pp. 35-36.

[59] S. 1297 and S. 1285 differed with respect to the roles and authorities assigned to the director and board. S. 1297 gave most of the power to the director (with the board in an advisory capacity); while S. 1285 put most of the authority in the hands of the board, who appointed the director.

[60] U.S. Congress, House Committee on Science and Technology, Task Force on Science Policy, "A History of Science Policy in the United States, 1940-1985," *Science Policy Study Background Report No. 1*, 99th Cong., 2nd sess., September 1986 (Washington, DC: GPO, 1986), p. 21-27; and, J. Merton England, *A Patron for Pure Science: The National Science Foundation's Formative Years, 1945-57* (Washington, DC: NSF, 1982).

As drafted in August of 1945, S. 1297 and S. 1285 would have resolved these policy issues differently. Senator Kilgore's bill (S. 1297) envisioned a scientific foundation that was administered by a publically appointed director and advised by a board, that distributed funding and research findings broadly, and that defined the term "research and development" to include both theoretical exploration as well as the extension of investigation

> into practical application, including the preparation of plans, specifications, and standards for various goods and services, the undertaking of related economic and industrial studies, the experimental production and testing of models, and the building and operation of pilot plants.[61]

Senator Magnuson's bill (S. 1285), on the other hand, would have created a research foundation led by a publically appointed board that would select, direct, and supervise a director. The powers and duties of the foundation as described in S. 1285 include developing national science policies and support of basic research in the fields of mathematical, physical, and biological sciences. The bill does not include provisions for the broad distribution of funding, though it does authorize the publication and dissemination of research findings.

The differences between these approaches were not resolved in the 79th Congress. However, after two more years of debate Congress presented a bill to establish a National Science Foundation to President Harry S. Truman on July 25, 1947 (S. 526, National Science Foundation Act of 1947). Truman vetoed the bill. In his veto message Truman expressed two concerns. First, Truman asserted that S. 526 violated his appointment powers and raised questions about accountability because it did not provide for a presidentially appointed director. (S. 526 gave authority to appoint a director to the foundation.) Second, the President expressed conflict-of-interest concerns. As defined in S. 526, the foundation included 24 eminent scientists appointed by the President with the advice and consent of the Senate. These 24 scientists would determine who would receive foundation grants, which Truman perceived as a conflict of interest that "would inevitably give rise to suspicions of favoritism."[62]

In April 1950, Congress sent the President a new bill, S. 247 (National Science Foundation Act of 1950).[63] President Truman signed S. 247, which became P.L. 81-507 (hereinafter referred to as NSF's "organic act") on May 10, 1950.[64] NSF's organic act provided for an independent federal agency administered by a presidentially appointed board *and* director. As established in its organic act, NSF was empowered to develop and encourage a national policy for the promotion of basic research and science education, to support basic research in the mathematical, physical, medical, biological, engineering, and "other" (e.g., social) sciences. Section 3(b) addressed the geographic distribution issue obliquely, stating that it

---

[61] S. 1297, Title IV, Section 402 (a) as published in U.S. Senate, Committee on Military Affairs, Subcommittee on War Mobilization, *Legislative Proposals for the Promotion of Science: The Texts of Five Bills and Excerpts from Reports*, subcommittee print, 79th Cong., 1st sess., August 1945.

[62] Harry S. Truman Library & Museum, Public Papers of the Presidents, Harry S. Truman 1945-1953, "169. Memorandum of Disapproval of the National Science Foundation Bill," August 6, 1947, at http://www.trumanlibrary.org/publicpapers/index.php?pid=1918.

[63] S. 247 (National Science Foundation Act of 1950).

[64] Harry S. Truman Library & Museum, Public Papers of the Presidents, Harry S. Truman 1945-1953, "120. Statement by the President Upon Signing Bill Creating the National Science Foundation," May 10, 1950, at http://trumanlibrary.org/publicpapers/index.php?pid=743..

shall be one of the objectives of the Foundation to strengthen basic research and education in the sciences, including independent research by individuals, throughout the United States, including its Territories and possessions, and to avoid undue concentration of such research and education.[65]

As with prior versions of the bill, NSF's organic act specifically authorized the foundation to provide for scholarships and fellowships, to foster information exchange among scientists in the U.S. and abroad, to establish commissions, to act as a central clearinghouse for information about scientific and technological personnel, and to establish research divisions. With respect to the patent issues, P.L. 81-507 left these questions to the NSF to decide through the contract process.[66] With one notable exception, Congress did not pass another NSF authorization act for the next 15 years. (NSF's organic act provided the foundation with $500,000 in FY1951 and $15,000,000 annually thereafter. Congress amended the act in 1953 to provide "such sums as may be necessary" (P.L. 83-223).)

The next major reauthorization of the NSF organic act came in 1968.[67] In 1965, the House Committee on Science and Astronautics, Subcommittee on Science, Research and Development (chaired by Representative Emilio Daddario) undertook an extensive, three-year examination of the foundation's activities and legal authority. Some historians assert that renewed interest in the NSF organic act stemmed from concern about U.S. science policy post-*Sputnik*.[68] The result of the Daddario committee's work was P.L. 90-407 (An Act to Amend the National Science Foundation Act of 1950). P.L. 90-407 made several critical changes to the NSF organic act that harkened back to the establishment debates of the 1940s. In particular, the act expressly authorized NSF activities in the social sciences and it specifically authorized support for applied research.

P.L. 90-407 also changed NSF's authorization cycle. The 1968 act repealed the indefinite authorization established by P.L. 83-223 in 1953 and replaced it with an annual authorization. The one-year authorization cycle established by P.L. 90-407 was in place (generally) from FY1969 until FY1989. It was not unchallenged, however. During the late 1970s and early 1980s Congress debated whether to maintain the one-year authorization cycle for NSF. Some Members of Congress preferred tighter oversight and control over the foundation and therefore argued for the one-year authorization.[69] Other Members asserted that longer authorization cycles would assist in

---

[65] P.L. 81-507, Section 3(b).

[66] For a broader treatment of federal patent issues, please see CRS Report R42014, *The Leahy-Smith America Invents Act: Innovation Issues*, by John R. Thomas.

[67] Although not a reauthorization act per se, in 1962 President John F. Kennedy signed "Reorganization Plan No. 2 of 1962," which established the Office of Science and Technology (OST) within the Executive Office of the President. The plan transferred authority for national science policy making from NSF to OST and made other changes within NSF. Congress had the power to disapprove of this plan, but did not do so, and thereby facilitated implementation. (For more information about the reorganization process, see CRS Report R42852, *Presidential Reorganization Authority: History, Recent Initiatives, and Options for Congress*, by Henry B. Hogue.)

[68] Many analysts and historians consider the Soviet Union's launch of *Sputnik*, the world's first artificial satellite, a watershed moment in U.S. science (and science education) policy history. See CRS Report RL34263, *U.S. Civilian Space Policy Priorities: Reflections 50 Years After Sputnik*, by Deborah D. Stine.

[69] Ken Hechler, *Toward the Endless Frontier: History of the Committee on Science and Technology, 1959-79* (Washington, D.C.: U.S. House of Representatives/GPO, 1980), pp. 537-538.

long-range planning, ensure stable funding, and facilitate "sound national science policy and programs."[70] These legislators typically argued for at least two-year authorizations.

Since FY1989 NSF authorization cycles have generally extended beyond a single year. Enacted authorizations for the NSF over the past two decades have typically fluctuated between three and five years. (See **Table A-1**.)

America COMPETES

Since 2007, Congress has included language to reauthorize the NSF in broader bills that, among other things, also authorized scientific research at the Department of Energy's Office of Science and the National Institute of Standards and Technology. Known colloquially as the "COMPETES acts,"[71] these measures authorized FY2008 through FY2013 funding levels for selected federal research accounts, authorized certain federal STEM education programs, and addressed various other policy issues associated with innovation and national competitiveness. NSF provisions in the 2007 and 2010 COMPETES acts included funding authorizations for most major foundation accounts as well as policy provisions authorizing or amending specified policies and programs related to research, STEM education, and broadening participation. Legislators in the 113th Congress have begun considering the reauthorization of major provisions from the COMPETES acts.

Doubling path. A primary policy question facing the next NSF reauthorization is whether to continue authorizing funding increases for NSF as part of the COMPETES acts "doubling path" policy. Under this policy, Congress and two successive Administrations sought to double—over several years—combined funding for certain federal accounts[72] (including NSF) that fund substantial levels of physical sciences and engineering (PS&E) research.[73] PS&E research is widely believed to contribute to U.S. economic growth and national security by creating the underlying knowledge that supports technological innovation. The COMPETES-authorized PS&E doubling effort followed a successful effort to double funding for medical research at the National Institutes of Health.[74]

As enacted in the 2007 America COMPETS Act, funding authorizations for the targeted accounts set a pace for doubling over approximately seven years (as compared to the FY2006 baseline). However, actual appropriations followed about an 11-year doubling pace during the FY2008 to FY2010 authorization period. The America COMPETES Reauthorization Act of 2010, which authorized funding for the targeted accounts from FY2011 through FY2013, established a doubling pace of about 11 years. (That is, the reauthorization followed a growth rate that was consistent with the growth in actual appropriations during the first authorization period.)

---

[70] S.Rept. 95-851, pp. 22-23.

[71] America COMPETES Act (P.L. 110-69) and America COMPETES Reauthorization Act of 2010 (P.L. 111-358).

[72] The targeted accounts included the NSF, the Department of Energy's Office of Science, and the Scientific and Technical Research and Services (STRS) and Construction of Research Facilities (CRF) accounts at the National Institute of Standards and Technology (NIST).

[73] The largest funder of research in engineering is the Department of Defense. The National Aeronautics and Space Administration (NASA) also emphasizes engineering and the physical sciences research. See National Science Board, *Science and Engineering Indicators 2014*, NSB-14-01 (Arlington, VA: National Science Foundation, 2014), p. 4-38.

[74] For more information about the NIH doubling, see CRS Report R43341, *A History of NIH Funding*, by Judith A. Johnson.

Appropriations to the targeted accounts decreased from the prior year in both FY2011 and FY2013, and as a result, increased the doubling timeframe set by actual appropriations from about 11 years to about 18 years.[75]

The idea of an NSF budget doubling did not originate with the COMPETES acts. President Ronald Reagan proposed a five-year doubling of the NSF budget in January 1987.[76] His FY1988 and FY1989 budget requests sought increases that were consistent with this approach. In October 1988, Congress enacted funding authorizations that sought to double NSF's budget in approximately five years as part of P.L. 100-570 (National Science Foundation Authorization Act of 1988). Actual appropriations to the NSF increased by about 59% during this period.[77] In 2002, Congress passed and President George W. Bush signed, P.L. 107-368 (National Science Foundation Authorization Act of 2002). P.L. 107-368 authorized increases in the NSF budget that were consistent with a five-year doubling. However, the Bush Administration is reported to have objected to the notion of doubling as an arbitrary goal for the NSF, and language referring to doubling was removed from the final bill, though the authorization increases remained.[78] Actual appropriations to the NSF increased by about 22% during the P.L. 107-368 authorization period.[79] President Bush later proposed a doubling similar to that authorized by the COMPETES acts—for example, focused on the targeted accounts, not just NSF—in the 2006 American Competitiveness Initiative.[80]

Viewed in appropriations by decade (e.g., FY1951 to FY1960, FY1960 to FY1970, etc.), the NSF budget doubled (in current dollars) over the course of each of the five decades between the foundation's first budget in FY1951 and FY1990.[81] (See **Table 1**.) Growth slowed from this pace around the turn of the 21st century. Between FY1990 and FY2000, the NSF budget grew by about 88% in current dollars; between FY2000 and FY2010, it grew by about 76% in current dollars.[82]

In inflation-adjusted (constant) dollars, NSF's budget doubled between FY1951 and FY1960, and again between FY1960 and FY1970. The NSF budget has not doubled by decade (in constant dollars) since then. Between FY1970 and FY1980, NSF's budget grew at its lowest constant dollar rate (16%). Between FY1980 and FY2010, NSF constant dollar funding increased by 38% or more each decade. However, constant dollar funding for NSF was below FY2010 levels in

---

[75] For more information about the PS&E doubling effort, see CRS Report R41951, *An Analysis of Efforts to Double Federal Funding for Physical Sciences and Engineering Research*, by John F. Sargent Jr.

[76] President Ronald Reagan, "Radio Address to the Nation on Administration Goals," radio address, January 31, 1987, at http://www.presidency.ucsb.edu/ws/index.php?pid=34674.

[77] NSF received $1.717 billion in appropriations in FY1988. P.L. 100-570 authorized NSF funding increases from FY1989 ($2.050 billion) through FY1993 ($3.505 billion). Actual appropriations to NSF in FY1993 were $2.734 billion, or $1.017 million (59%) more than the FY1988 funding level.

[78] Jeffrey Mervis, "Bush Signs NSF 'Doubling' Bill," *Science*, December 20, 2002, at http://news.sciencemag.org/2002/12/bush-signs-nsf-doubling-bill.

[79] NSF received $4.823 billion in appropriations in FY2002. P.L. 107-368 authorized NSF funding increases from FY2003 ($5.536 billion) to FY2007 ($9.839 billion). Actual appropriations to NSF in FY2007 were $5.890 billion, or $1.067 million (22%) more than the FY2002 funding level.

[80] Executive Office of the President, Domestic Policy Council, Office of Science and Technology Policy, *American Competitiveness Initiative: Leading the World in Innovation*, February 2006, at http://georgewbush-whitehouse.archives.gov/stateoftheunion/2006/aci/aci06-booklet.pdf.

[81] Other periods of time or funding units might produce different results. The decade-long perspective is largely consistent with the 11-year doubling pace enacted in the America COMPETES Reauthorization Act of 2010 (P.L. 111-358).

[82] This growth estimate excludes American Recovery and Reinvestment Act (ARRA, P.L. 111-5) funding.

FY2011, FY2012, FY2013, and FY2014, which shows that funding for NSF has not kept pace with inflation this decade.

Table 1. NSF Appropriations by Decade: FY1951 to FY2010

In Millions, Current and Constant (FY2015) Dollars, Rounded

Year	Current ($ millions)	Constant (FY2015 $ millions)
FY1951	0	2
FY1960	153	963
FY1970	440	2,163
FY1980	992	2,500
FY1990	2,082	3,445
FY2000	3,912	5,266
FY2010	6,873	7,471

Source: Excerpted from **Table B-1**.

Should Congress continue to pursue the COMPETES doubling policy in FY2014 and beyond? Many advocates assert that federal funding for PS&E basic research is inadequate (particularly in light of other countries' investments in R&D) and that more investment is needed to assure U.S. national security and competitiveness.[83] Continuing to provide increased authorizations for the targeted accounts at some (to be determined) doubling rate might signal Congress's continued commitment to these accounts and to the doubling path policy. Further, some analysts have argued that the PS&E doubling policy, although not fully realized, may have protected the targeted accounts from reductions or slower growth during a period of constrained resources.[84] On the other hand, it may be challenging for the scientific community to plan for large or long-term projects without a clearer signal from Congress as to the actual budgetary resources they might receive.[85] Some policy makers who seek general reductions in federal expenditures may

---

[83] This case is laid out more fully in National Academy of Sciences, National Academy of Engineering, and Institute of Medicine, Committee on Prospering in the Global Economy of the 21st Century: An Agenda for America Science and Technology, and Committee on Science, Engineering, and Public Policy, *Rising Above the Gathering Storm: Energizing and Employing America for a Brighter Economic Future*, National Academies Press, 2007, http://www.nap.edu/catalog/11463 html.

[84] Testimony of Boston University Associate Professor of Strategy and Innovation and National Bureau of Economic Research, Research Associate Dr. Jeffrey L. Furman, in U.S. Congress, Senate Committee on Commerce, Science, and Transportation, *Five Years of the America COMPETES Act: Progress, Challenges, and Next Steps*, hearings, 112th Cong., 2nd sess., September 19, 2012, at http://www.commerce.senate.gov/public/?a=Files.Serve&File_id=8687a045-afdc-4b74-ac75-efca96893a88.

[85] For example, a 2012 *Science* magazine report noted that the U.S. astronomical community had to revisit the priorities laid out in the 2010 Astronomy and Astrophysics Decadal Survey, which was drafted under the assumption of an NSF doubling, after it became clear that actual appropriations were not keeping pace with COMPETES act-authorized funding levels. See Yudhijit Bhattacharjee, "Panel Says NSF Should Shutter Six U.S. Instruments," *Science*, vol. 337, August 24, 2012, http://www.sciencemag.org/content/337/6097/899.summary. Additionally, a September 2013 *Nature* editorial asserts that, at least in part due to the signal policy makers sent with the 2007 COMPETES act, NSF committed to two large ocean science division construction projects. Once completed, those projects will require operational support. *Nature* asserts that these operating costs will increase budget pressure on NSF's ocean science research account and that NSF should have anticipated that "big budgets would not last." See "Counting the Cost," *Nature* editorial, September 25, 2013, at http://www.nature.com/news/counting-the-cost-1.13804.

object to policies that seek to increase federal spending. However, other observers describe federal funding for basic research (like that funded by the NSF) as the "seed corn" that supports the U.S. economy.[86] These analysts may perceive NSF funding as a vital investment. Another view holds that federal funding for scientific research should be continued, but asserts that dollars should be focused "where links between science and application are well established, to deliver short- to medium-term benefits" rather than on the types of research NSF typically supports, which may be perceived by some observers as less targeted or less immediately commercially relevant.[87]

STEM education. Several inventories of the federal STEM education effort have highlighted NSF's important role—both in terms of funding and in the number and breadth of programs—in the federal STEM education portfolio. The NSF is the only federal agency whose primary mission includes supporting education across all fields of science and engineering. As such, funding for STEM education at the NSF impacts not only the agency, but also the entire federal STEM education effort.[88]

The COMPETES acts authorized increased funding for NSF's main education account, Education and Human Resources (EHR), and made various changes to specified NSF STEM education programs. Actual appropriations to EHR have not typically reached COMPETES-authorized levels. Further, Congress reduced enacted funding levels (from the prior year) for EHR in both FY2011 and FY2012. These reductions followed several years of fluctuating funding, as well as changes in the distribution of the foundation budget that reduced funding for EHR as a percentage of the total NSF budget. FY2013 actual funding levels for EHR were close to FY2012 levels.[89]

In addition to funding authorizations, the COMPETES acts authorized and amended some NSF STEM education programs.[90] Among the amended programs were the Graduate Research Fellowship (GRF) program and the Integrative Graduate Research and Education Traineeship (IGERT). First, Section 510 of the America COMPETES Reauthorization Act of 2010 sought to require NSF to treat the GRF and IGERT equally by increasing or decreasing funding for these programs at the same rates. This does not appear to have been implemented. Funding for the IGERT was reduced from the prior year each fiscal year between FY2011 and FY2013 while funding for the GRF increased each year. Second, Section 510 also directed NSF to draw from the RRA account for at least half of the funding it provides to the GRF and IGERT programs. RRA funding for the GRF and IGERT programs was close to 50% in FY2012 and FY2013.

The GRF was established in 1951 and is one of the oldest and most prestigious federal graduate research fellowships. GRF fellows receive a three-year, portable stipend of $32,000 annually and

---

[86] Jules Duga and Tim Studt. "Government Spending Continues to Drive R&D Growth." *R&D,* vol. 47, no. 1 (2005), pp. F3-F7,F10-F15.

[87] Daniel Sarewitz, "Double Trouble? To Throw Cash at Science Is a Mistake." *Nature,* vol. 468, no. 135 (November 10, 2010), http://www.nature.com/news/2010/101110/full/468135a.html; and Daniel Sarewitz, "Blue-Sky Bias Should Be Brought down to Earth," *Nature,* vol. 481, no. 7339 (January 4, 2012), at http://www.nature.com/news/blue-sky-bias-should-be-brought-down-to-earth-1.9722.

[88] For more information about the federal STEM education effort, see CRS Report R42642, *Science, Technology, Engineering, and Mathematics (STEM) Education: A Primer,* by Heather B. Gonzalez and Jeffrey J. Kuenzi.

[89] For more information about STEM education funding at NSF, see CRS Report R42470, *An Analysis of STEM Education Funding at the NSF: Trends and Policy Discussion,* by Heather B. Gonzalez.

[90] Most NSF STEM education programs are operated under general authority.

a $12,000 cost-of-education allowance for tuition and fees (paid to their institutions).[91] The IGERT, which began in 1997, is NSF's flagship interdisciplinary training program.[92] IGERT funding is awarded to institutions of higher education, which may utilize IGERT funding for student support or education research. In FY2013, NSF provided funding for 5,758 GRF fellows[93] and 1,572 IGERT trainees.

After the expiration of major provisions in the 2010 COMPETES reauthorization, President Obama proposed a major restructuring of federal STEM education programs as part of his FY2014 budget request. Under the proposed reorganization, NSF would have become the lead federal agency for undergraduate education and federal fellowships. For a range of reasons, congressional appropriators largely rejected the plan.[94] The Obama Administration released a "fresh" reorganization plan as part of the FY2015 budget request.[95] Congressional debate about the reorganization of the federal STEM education effort is ongoing.

For many policy makers, the prospect of a reorganized federal STEM education effort raises the question, "to what end?" The America COMPETES Reauthorization Act of 2010 directed the National Science and Technology Council (NSTC) to develop a strategy for federal STEM education programs, including those at NSF. That strategy was published in May 2013, after release of the Administration's FY2014 proposed reorganization of federal STEM education programs.[96] Some policy makers perceived the strategy as insufficiently independent from the FY2014 proposed reorganization;[97] while others perceived it as a starting place for a new conversation about the federal STEM education portfolio in lieu of the proposed reorganization.[98] Policy makers' interest in the form and function of the federal STEM education portfolio, as well as the character of NSF's role within that portfolio, continues.

---

[91] National Science Foundation, Graduate Research Fellowship Program, "About the NSG Graduate Research Fellowship Program," National Science Foundation Website, accessed on February 20, 2014, at http://www.nsfgrfp.org/about_the_program. NSF's *FY2015 Budget Request to Congress* seeks to increase the GRF stipend to $34,000. (Available at http://www.nsf.gov/about/budget/fy2015/index.jsp.)

[92] National Science Foundation, Integrative Graduate Education and Research Traineeship, "Introduction to the IGERT Program," *National Science Foundation Website*, accessed on February 20, 2014, at http://www.nsf.gov/crssprgm/igert/intro.jsp.

[93] The GRF program total includes fellows across multiple cohorts. On average, the GRF program issues about 2,000 *new* fellowships each year. E-mail communication between CRS and staff from the NSF Office of Legislative and Public Affairs, dated February 21, 2014.

[94] Different legislators rejected the proposal—which included changes across a wide variety of programs and agencies—for different reasons. The "Joint Explanatory Statement" published in the January 15, 2014, *Congressional Record*, which accompanied P.L. 113-76 (Consolidated Appropriations Act, 2014), asserts that "the proposal contained no clearly defined implementation plan, had no buy-in from the education community, and failed to sufficiently recognize or support a number of proven, successful programs."

[95] For more information about the proposed FY2015 STEM education reorganization, see CRS Report IF00013, *The President's FY2015 Budget and STEM Education (In Focus)*, by Heather B. Gonzalez; and CRS Report IN10011, *The Administration's Proposed STEM Education Reorganization: Where Are We Now?*, by Heather B. Gonzalez.

[96] Among other things, Section 101 of P.L. 111-358 directs the NSTC to develop a strategy for federal STEM education programs. NSTC published this strategy in June 2013. See Executive Office of the President, National Science and Technology Council, Committee on STEM Education, *Federal Science, Technology, Engineering, and Mathematics (STEM) Education: 5-Year Strategic Plan*, May 2013, available at http://www.whitehouse.gov/sites/default/files/microsites/ostp/stem_stratplan_2013.pdf.

[97] H.Rept. 113-171, p. 59.

[98] House Committee on Science, Space, and Technology, Ranking Member Eddie Bernice Johnson, "Committee Discusses Proposed Reorganization of STEM Education Programs," press release, June 4, 2013.

Reauthorization Activity in the 113th Congress

Major provisions of the COMPETES acts, including provisions authorizing funding for the National Science Foundation, expired in FY2013. The House Committee on Science, Space, and Technology has begun the process of reauthorizing certain COMPETES provisions. On March 13, 2014, the Subcommittee on Research and Technology marked up and approved by voice vote the Frontiers in Innovation, Research, Science, and Technology (FIRST) Act (H.R. 4186).[99] The full committee marked up the FIRST Act on May 21, 2014, and voted on amendments on May 28, 2014. The ranking Member of the House Committee on Science, Space, and Technology has introduced an alternative bill, H.R. 4159 (America COMPETES Reauthorization Act of 2014). H.R. 4159 had not been marked up as of the date of this report. H.R. 4186 and H.R. 4159 differ from each other, and from the COMPETES acts, in significant ways. These differences, however, are beyond the scope of this report.

Budget and Appropriations

Like other federal agencies, NSF's annual budget requests to Congress provide insight into foundation activities and priorities.[100] A summary of NSF's three most recent budget requests (FY2013, FY2014, and FY2015), and associated appropriations activity, follows. **Table B-1** provides NSF authorizations, budget requests, and actual appropriations in current and constant (inflation-adjusted) dollars from FY1951 to FY2015. **Table B-2** provides FY2003 to FY2013 NSF obligations by major account.

NSF adopted its current appropriations account structure in FY2003. In general, NSF's major accounts have been comparable since then.[101] NSF has six major appropriations accounts: Research and Related Activities (RRA), Education and Human Resources (EHR), Major Research Equipment and Facilities Construction (MREFC), Agency Operations and Award Management (AOAM), National Science Board (NSB), and the Office of the Inspector General (OIG). The majority of NSF's primary mission activities are funded through RRA and EHR.

FY2015

For FY2015, the Obama Administration seeks $7.255 billion in funding for the NSF. This amount is $83 million (1.2%) over the FY2014 enacted level of $7.172 billion. The request holds funding levels for RRA and MREFC essentially constant while seeking a 5.1% increase ($43 million) for EHR as well as a 13.5% increase ($40 million) for AOAM. Most of the new AOAM funding would apply toward a new NSF headquarters. NSF's FY2015 budget request to Congress highlights five initiatives that were also foundation priorities in FY2014 and FY2013: Cognitive Science and Neuroscience ($29 million); Cyber-enabled Materials, Manufacturing, and Smart

---

[99] A webcast of the mark-up, as well as related hearing materials (including amendments and vote tallies), is available at http://science.house.gov/markup/subcommittee-research-and-technology-markup-hr-4186.

[100] NSF publishes its annual budget requests to Congress—dating from FY1998 to FY2015—on its website at http://www.nsf.gov/about/budget/. Additionally, policy makers may access award summaries by state and institution, historical NSF account data, and related reports at http://dellweb.bfa nsf.gov/.

[101] NSF changed its account structure in FY2003. Prior years are not comparable. In FY2008, NSF shifted the EPSCoR program from EHR to RRA. **Table B-2** treats EPSCoR as an RRA sub-account for all years in the data set.

Systems ($213 million); Cyberinfrastructure Framework for 21st Century Science, Engineering, and Education ($125 million); Science, Engineering, and Education for Sustainability ($139 million); and Secure and Trustworthy Cyberspace ($100 million).[102] The FY2015 NSF budget request incorporates STEM education programs changes in accordance with the Administration's revised FY2015 government-wide reorganization of federal STEM education programs.[103] In addition to NSF's regular budget request, the Administration seeks $552 million in funding for NSF through the proposed Opportunity, Growth, and Security Initiative.[104]

Congress had not enacted legislation authorizing FY2015 funding levels for NSF when the President released the FY2015 budget request.

FY2014

FY2014 enacted funding for NSF is $7.172 billion. This amount is $270 million (3.9%) more than NSF's FY2013 actual funding level of $6.902 billion. Most of the $270 million increase ($250 million) went to RRA. FY2014 enacted funding for NSF's six major accounts is $5.809 billion for RRA (including $158 million for EPSCoR), $847 million for EHR, $200 million for MREFC, $298 million for AOAM, $4 million for NSB, and $14 million for OIG.[105] Congress had not enacted legislation authorizing funding for the NSF in FY2014 when FY2014 appropriations were enacted.

The Obama Administration initially sought $7.626 billion in funding for the NSF in FY2014. NSF's FY2014 budget request to Congress noted that its overarching priorities for FY2014 would include six programs: Cyber-enabled Materials, Manufacturing, and Smart Systems; Cyberinfrastructure Framework for 21st Century Science, Engineering, and Education; NSF Innovation Corps; Integrated NSF Support Promoting Interdisciplinary Research and Education; Science, Engineering, and Education for Sustainability; and Secure and Trustworthy Cyberspace.[106] The FY2014 NSF budget request also incorporated several changes to the foundation's STEM education programs in accordance with the Administration's proposed FY2014 government-wide reorganization of federal STEM education programs.[107]

The House and Senate Committees on Appropriations recommended $6.995 billion and $7.426 billion, respectively, for NSF in FY2014. Both committees initially rejected the Administration's proposed changes to the federal STEM education effort, including changes to NSF programs. The final FY2014 appropriations agreement reiterated this objection. The appropriations committees

---

[102] With one exception—the Expeditions in Education program, which was not included in the FY2014 request—these were the same programs included in the "OneNSF Framework" from NSF's FY2013 budget request.

[103] National Science Foundation, *FY2014 Budget Request to Congress*, April 10, 2013, http://www.nsf.gov/about/budget/fy2014/index.jsp.

[104] Information about this initiative is included in the President's FY2015 budget request, available at http://www.whitehouse.gov/sites/default/files/omb/budget/fy2015/assets/budget.pdf.

[105] For more information about the NSF FY2014 budget request and appropriations, see CRS Report R43080, *Commerce, Justice, Science, and Related Agencies: FY2014 Appropriations*, coordinated by Nathan James, Jennifer D. Williams, and John F. Sargent Jr.

[106] With one exception—the Expeditions in Education program, which was not included in the FY2014 request—these were the same programs included in the "OneNSF Framework" from NSF's FY2013 budget request.

[107] National Science Foundation, *FY2014 Budget Request to Congress*, April 10, 2013, http://www.nsf.gov/about/budget/fy2014/index.jsp.

initially disagreed on funding for the Large Synoptic Survey Telescope (LSST) in the MREFC account—the Senate Committee on Appropriations sought to fund the new project, the House Committee on Appropriations would not. The final agreement provided some of the requested funding for the LSST and encouraged the foundation to seek permission to transfer funds from other accounts if the amount appropriated was insufficient.[108]

FY2013

FY2013 actual funding for NSF was $6.902 billion. This amount was $199.4 million (3.5%) less than NSF's FY2012 actual funding level of $7.105 billion.[109] FY2013 actual funding levels for NSF's six major accounts were $5.559 billion (RRA), $835 million (EHR), $197 million (MREFC), $294 million (AOAM), $4 million (NSB), and $14 million (OIG). The America COMPETES Reauthorization Act of 2010 (P.L. 111-358) authorized $8.300 billion in funding at the NSF in FY2013.

The Obama Administration sought $7.373 billion for the NSF in FY2013. NSF's FY2013 budget request sought to continue increasing NSF's budget in accordance with the doubling path policy and emphasized "OneNSF Framework" priorities. NSF's FY2013 budget documents asserted that the six OneNSF Framework priority programs would "create new knowledge, stimulate discovery, address complex societal problems, and promote national prosperity."[110]

The full House and Senate Committee on Appropriations agreed on identical funding levels for five of six major NSF accounts in FY2013. The primary difference between the two proposals was in the main research account (RRA). The House proposed $5.943 billion for RRA in FY2013 while the Senate Committee on Appropriations recommended $5.883 billion. This difference in funding for RRA lead to an equivalent difference in House and Senate topline recommendations for NSF, which were $7.333 billion and $7.273 billion, respectively. FY2013 House and Senate recommendations for NSF's other five major accounts were $875.6 million (EHR), $196.2 million (MREFC), $299.4 million (AOAM), $4.4 million (NSB), and $14.2 million (OIG). These amounts equaled the Administration's requested funding levels.

Concluding Observations

The National Science Foundation plays a key role in the federal research and development, as well as STEM education, portfolios. It is a primary source of support for basic research in fields that many analysts cite as key to future competitiveness, such as the physical sciences, mathematics, and computer science. It is also a primary source for federal STEM education funding. Yet, the foundation differs from many other federal agencies in a number of key ways. It was established outside of the central core of executive agencies rather than under the direct control of the President. It focuses on fundamental research across a variety of scientific and

---

[108] H.Rept. 113-171, S.Rept. 113-78, and the Joint Explanatory Statement published in the January 15, 2014, *Congressional Record*.

[109] For more information about the NSF FY2013 budget request and appropriations, see CRS Report R42440, *Commerce, Justice, Science, and Related Agencies: FY2013 Appropriations*, coordinated by Nathan James, Jennifer D. Williams, and John F. Sargent Jr.

[110] National Science Foundation, *FY2013 Budget Request to Congress*, February 13, 2012, p. Overview-3, http://www.nsf.gov/about/budget/fy2013/index.jsp.

technological fields rather than on the specific mission needs of the federal government; and it is
the only federal agency whose primary mission includes STEM education. These differences
underpin much of the policy conversation about the NSF; and, as a practical matter, offer both
benefits and barriers to legislators seeking to apply NSF's various assets toward specific national
goals.

Appendix A. NSF Authorizations Acts

Table A-1. Selected NSF Authorizations Acts
FY1951-FY2013

Public Law	Bill Number	From	To
P.L. 81-507	S. 247	FY1951	FY1952
P.L. 83-223	S. 977	FY1953	not defined
authorizations not defined		FY1954	FY1968
P.L. 90-407	H.R. 5404	FY1969	FY1969
P.L. 91-120	S. 1857	FY1970	FY1970
P.L. 91-356	H.R. 16595	FY1971	FY1971
P.L. 92-86	H.R. 7960	FY1972	FY1972
P.L. 92-372	H.R. 14108	FY1973	FY1973
P.L. 93-96	H.R. 8510	FY1974	FY1974
P.L. 93-413	H.R. 13999	FY1975	FY1975
P.L. 94-86	H.R. 4723	FY1976	FY1976
P.L. 94-471	H.R. 12566	FY1977	FY1977
P.L. 95-99	H.R. 4991	FY1978	FY1978
P.L. 95-434	H.R. 11400	FY1979	FY1979
P.L. 96-44	H.R. 2729	FY1980	FY1980
P.L. 96-516	S. 568	FY1981	FY1981
some authorizations are introduced, none become law		FY1982	FY1985
P.L. 99-159	H.R. 1210	FY1986	FY1986
P.L. 99-383	H.R. 4184	FY1987	FY1987
some authorizations are introduced, none become law		FY1988	FY1988
P.L. 100-570	H.R. 4418	FY1989	FY1993
some authorizations are introduced, none become law		FY1994	FY1997
P.L. 105-207	H.R. 1273	FY1998	FY2000
some authorizations are introduced, none become law		FY2001	FY2001
P.L. 107-368	H.R. 4664	FY2003	FY2007
P.L. 110-69	H.R. 2272	FY2008	FY2010
P.L. 111-358	H.R. 5116	FY2011	FY2013

Source: Congressional Research Service, based on information from the Legislative Information System and Proquest Congressional.

Notes: This table includes a list of major NSF authorization acts as per a CRS search of historical legislative databases. The list of authorization acts has been reviewed by National Science Board legal counsel, who affirmed its apparent completeness. In addition to the above-listed authorization acts, other laws have also amended various parts of the NSF code.

Appendix B. NSF Funding History

Table B-1. NSF Authorizations, Budget Requests, and Appropriations
FY1951 to FY2015
In Millions, Current and Constant (FY2015) Dollars, Rounded

Fiscal Year	Current ($ millions)			Constant (FY2015 $ millions)		
	Authorization	Request	Appropriation	Authorization	Request	Appropriations
1951	such sums	-	0	n/a	-	2
1952	such sums	14	4	n/a	103	26
1953	such sums	15	5	n/a	109	34
1954	such sums	15	8	n/a	108	57
1955	such sums	14	14	n/a	100	101
1956	such sums	31	53	n/a	215	367
1957	such sums	41	40	n/a	276	267
1958	such sums	65	52	n/a	422	336
1959	such sums	140	138	n/a	894	878
1960	such sums	160	153	n/a	1,010	963
1961	such sums	190	176	n/a	1,180	1,092
1962	such sums	210	263	n/a	1,292	1,619
1963	such sums	358	323	n/a	2,175	1,959
1964	such sums	589	353	n/a	3,535	2,118
1965	such sums	488	420	n/a	2,877	2,479
1966	such sums	530	480	n/a	3,061	2,771
1967	such sums	525	481	n/a	2,942	2,695
1968	such sums	526	495	n/a	2,850	2,682
1969	525	500	400	2,719	2,590	2,072
1970	478	500	440	2,348	2,458	2,163
1971	538	513	513	2,516	2,400	2,400
1972	653	622	622	2,913	2,777	2,777
1973	697	653	649	2,982	2,795	2,778
1974	633	583	579	2,528	2,328	2,315
1975	808	672	764	2,925	2,435	2,768
1976	787	755	715	2,666	2,559	2,423
1977	811	802	776	2,562	2,535	2,452
1978	879	944	863	2,603	2,795	2,555
1979	930	934	911	2,549	2,560	2,497

Fiscal Year	Current ($ millions)			Constant (FY2015 $ millions)		
	Authorization	Request	Appropriation	Authorization	Request	Appropriations
1980	1,002	1,006	992	2,525	2,536	2,500
1981	1,115	1,148	1,025	2,559	2,636	2,354
1982	n/a	1,354	1,039	n/a	2,909	2,232
1983	n/a	1,073	1,094	n/a	2,208	2,251
1984	n/a	1,292	1,341	n/a	2,569	2,665
1985	n/a	1,502	1,502	n/a	2,890	2,889
1986	1,517	1,569	1,524	2,854	2,952	2,867
1987	1,685	1,686	1,623	3,101	3,102	2,986
1988	n/a	1,893	1,717	n/a	3,375	3,061
1989	2,050	2,050	1,923	3,515	3,515	3,296
1990	2,388	2,149	2,082	3,951	3,556	3,445
1991	2,782	2,485	2,316	4,445	3,971	3,701
1992	3,245	2,742	2,571	5,061	4,277	4,009
1993	3,505	3,037	2,734	5,340	4,627	4,165
1994	n/a	2,753	2,983	n/a	4,105	4,447
1995	n/a	3,200	3,264	n/a	4,672	4,765
1996	n/a	3,360	3,220	n/a	4,816	4,615
1997	n/a	3,325	3,270	n/a	4,683	4,606
1998	3,506	3,367	3,431	4,877	4,684	4,773
1999	3,773	3,773	3,676	5,184	5,184	5,050
2000	3,886	3,921	3,912	5,231	5,278	5,266
2001	n/a	4,572	4,431	n/a	6,009	5,823
2002	n/a	4,473	4,823	n/a	5,785	6,239
2003	5,536	5,036	5,323	7,027	6,391	6,756
2004	6,391	5,481	5,589	7,915	6,789	6,922
2005	7,378	5,745	5,482	8,860	6,899	6,584
2006	8,520	5,605	5,589	9,909	6,519	6,500
2007	9,839	6,020	5,890	11,142	6,817	6,670
2008	6,600	6,429	6,125	7,323	7,133	6,796
2009	7,326	6,854	6,494	8,034	7,516	7,121
2010	8,132	7,045	6,873	8,841	7,659	7,471
2011	7,424	7,424	6,806	7,917	7,917	7,257
2012	7,800	7,767	7,033	8,174	8,140	7,371
2013	8,300	7,373	6,884	8,569	7,612	7,107

Fiscal Year	Current ($ millions)			Constant (FY2015 $ millions)		
	Authorization	Request	Appropriation	Authorization	Request	Appropriations
2014	n/a	7,626	7,172	n/a	7,757	7,295
2015	n/a	7,255	-	n/a	7,255	-

Source: Funding data in the "Authorization" columns are from selected FY1951 to FY2013 NSF authorization acts, as provided in **Table A-1**. Funding data in the "Request" and "Appropriations" columns are from National Science Foundation, Budget Internet Information System, "NSF Requests and Appropriations History," NSF.gov, accessed March 19, 2014, http://dellweb.bfa.nsf.gov/NSFRqstAppropHist/NSFRequestsandAppropriationsHistory.pdf. To calculate constant dollars, CRS applied the Gross Domestic Product, Chained Price Index (adjusted to reflect FY2015 dollars) found in Office of Management and Budget, "Table 10.1," Historical Tables, accessed on May 6, 2014, available at http://www.whitehouse.gov/sites/default/files/omb/budget/fy2015/assets/hist10z1.xls.

Notes: Totals may not add due to rounding. As per communication between CRS and NSF dated March 20, 2014, the "appropriation" column shows funding provided in annual appropriations acts plus adjustments required in those acts, other laws, and committee reports, etc. Adjustments include rescissions, sequestration, funding transfers across NSF accounts, supplemental appropriations (not including American Recovery and Reinvestment Act, P.L. 111-5, funding in FY2009), and other changes. The resulting amounts most closely align with NSF's approved Current Plans.

Table B-2. NSF Obligations by Major Account: FY2003-FY2013

In Millions, Current Dollars, Rounded

Fiscal Year	RRA	EHR	MREFC	AOAM	NSB	OIG	NSF Total
2003	$4,144	$846	$179	$189	$3	$9	$5,369
2004	$4,388	$850	$184	$219	$2	$9	$5,652
2005	$4,328	$750	$165	$223	$4	$10	$5,481
2006	$4,449	$700	$234	$247	$4	$11	$5,646
2007	$4,758	$696	$166	$248	$4	$12	$5,884
2008	$4,853	$766	$167	$282	$4	$12	$6,084
2009	$5,152	$846	$161	$294	$4	$12	$6,469
2010	$5,615	$873	$166	$300	$4	$14	$6,972
2011	$5,608	$861	$125	$299	$4	$14	$6,913
2012	$5,758	$831	$198	$299	$4	$14	$7,105
2013	$5,559	$835	$196	$294	$4	$14	$6,902

Source: National Science Foundation annual budget requests to Congress from FY2005 to FY2015.

Notes: NSF adopted its current appropriations account structure in 2003. CRS adjusted FY2003 to FY2007 RRA and EHR obligations data to reflect the transfer of the EPSCoR program between these accounts in FY2008. This table treats EPSCoR as a research account program for all years in the data set. Does not include American Recovery and Reinvestment Act (ARRA, P.L. 111-5) funding.

Author Contact Information

Heather B. Gonzalez
Specialist in Science and Technology Policy
hgonzalez@crs.loc.gov, 7-1895